A Comparison of Two Fires:
The Westview Towers
North Bergen, New Jersey
and
The Council Towers Apartments
St. Louis, Missouri

Investigated by: John Lee Cook, Jr.

This is Report 119 of the Major Fires Investigation Project conducted by Varley-Campbell and Associates, Inc./TriData Corporation under contract EME-97-CO-0506 to the United States Fire Administration, Federal Emergency Management Agency.

Department of Homeland Security
United States Fire Administration
National Fire Data Center

U.S. Fire Administration Fire Investigations Program

The U.S. Fire Administration develops reports on selected major fires throughout the country. The fires usually involve multiple deaths or a large loss of property. But the primary criterion for deciding to do a report is whether it will result in significant "lessons learned." In some cases these lessons bring to light new knowledge about fire--the effect of building construction or contents, human behavior in fire, etc. In other cases, the lessons are not new but are serious enough to highlight once again, with yet another fire tragedy report. In some cases, special reports are developed to discuss events, drills, or new technologies which are of interest to the fire service.

The reports are sent to fire magazines and are distributed at National and Regional fire meetings. The International Association of Fire Chiefs assists the USFA in disseminating the findings throughout the fire service. On a continuing basis the reports are available on request from the USFA; announcements of their availability are published widely in fire journals and newsletters.

This body of work provides detailed information on the nature of the fire problem for policymakers who must decide on allocations of resources between fire and other pressing problems, and within the fire service to improve codes and code enforcement, training, public fire education, building technology, and other related areas.

The Fire Administration, which has no regulatory authority, sends an experienced fire investigator into a community after a major incident only after having conferred with the local fire authorities to insure that the assistance and presence of the USFA would be supportive and would in no way interfere with any review of the incident they are themselves conducting. The intent is not to arrive during the event or even immediately after, but rather after the dust settles, so that a complete and objective review of all the important aspects of the incident can be made. Local authorities review the USFA's report while it is in draft. The USFA investigator or team is available to local authorities should they wish to request technical assistance for their own investigation.

This report and its recommendations were developed by USFA staff and by Varley-Campbell & Associates, Inc. Miami and Chicago, its staff and consultants, who are under contract to assist the Fire Administration in carrying out the Fire Reports Program.

The United States Fire Administration greatly appreciates the cooperation received from the North Bergen, New Jersey and St. Louis, Missouri Fire Departments.

For additional copies of this report write to the U.S. Fire Administration, 16825 South Seton Avenue, Emmitsburg, Maryland 21727. The report is available on the Administration's Web site at http:// www.usfa.dhs.gov/

U.S. Fire Administration

Mission Statement

As an entity of the Department of Homeland Security, the mission of the USFA is to reduce life and economic losses due to fire and related emergencies, through leadership, advocacy, coordination, and support. We serve the Nation independently, in coordination with other Federal agencies, and in partnership with fire protection and emergency service communities. With a commitment to excellence, we provide public education, training, technology, and data initiatives.

 FEMA

TABLE OF CONTENTS

A Comparison of Two Fires:
The Westview Towers, North Bergen, New Jersey
and
The Council Towers Apartments
St. Louis, Missouri

OVERVIEW

On August 9, 1998, a fire occurred in an apartment located on the fourth floor of the Westview Towers Building in North Bergen, New Jersey. Four residents of the building died and thirty-two people, including seven emergency responders, were injured seriously enough to require transportation to a hospital. Twenty-two firefighters and an undetermined number of residents were also treated at the scene for minor injuries due to heat, smoke inhalation, minor burns, cuts, and bruises.

Two of the fatalities occurred in the apartment of origin, #4E, when the victims were trapped by heat and smoke conditions that forced them to retreat to the balcony of their apartment. The balcony was inaccessible by an aerial device and firefighters attempted to reach the trapped victims with ground ladders. One victim fell to her death during this attempt because she lunged at a ladder, which had not been secured to the balcony railing. The other victim succumbed to the intense heat and smoke conditions on the balcony before firefighters could reach her.

Two additional victims were discovered in a stairwell between the sixth and seventh floors. They were residents of the tenth floor and were overcome by smoke as they attempted to escape down the stairwell. Firefighters removed them to an apartment on the sixth floor, but were unsuccessful in their attempt to revive them.

It took more than eighty firefighters to extinguish the blaze in the twenty-two-story building. Companies from West New York, Weehawken, Jersey City, and Hoboken assisted the North Bergen Fire Department with the fire. The fire resistive construction of the building limited the spread of the fire, but contributed to the development of severe heat conditions on the fire floor and delayed extinguishment. The building was not sprinklered. The stairwells were not pressurized and this allowed the smoke to spread to the upper floors of the building.

After the fire, three oxygen bottles were discovered in Apartment 4E. An aluminum cylinder was found in the living room of the apartment and the cylinder ruptured during the fire. Two cylinders were found in the kitchen. The relief devices actuated during the fire, but the two cylinders remained intact. The release of the contents of the three cylinders increased the rate of combustion and resulted in the almost complete destruction of the contents in the room of origin. The intensity of the blaze also caused heavy spalling of the ceiling.

On October 12, 1998 in Saint Louis, Missouri a fire started in Apartment 2103 of the Council Tower Apartments Building and communicated out the windows to the apartment immediately above on the twenty-second floor. No one died in the eight-alarm blaze, but thirteen residents and three fire-fighters received sufficient injuries to require transportation to a hospital. Most of the injuries proved to be minor. However, a Fire Department Captain suffered severe burns to his respiratory tract and is not expected to be able to return to active duty as a firefighter due to extent of his injuries.

The extinguishment effort in the twenty-seven-story building required the response of over 150 firefighters and all but three of the department's fire companies. Mutual-aid companies were called in to fill vacant St. Louis fire stations and off-duty personnel also responded. The fire was similar to the fire in New Jersey because of the presence of three oxygen cylinders in the apartment of origin. At least two of the cylinders ruptured which intensified fire conditions and allowed heat and smoke to spread to an adjacent apartment on the fire floor. The Council Tower Apartments Building was also built of fire resistive construction and was not fully sprinklered.

The cause of both fires remains undetermined, but it is believed by fire officials in both jurisdictions that the fires were the result of smoking. The resident of the apartment of origin in the St. Louis fire smoked in spite of suffering from emphysema and chronic bronchitis. One of the two residents of the apartment of origin in North Bergen also smoked, is suffering from cancer, and having had one lung removed.

KEY ISSUES

Issues	Comments
Fire Building	Both fires occurred in highrise apartment buildings. The 27-story Council Tower Apartments Building was occupied exclusively by elderly residents. The 22-story Westview Towers Building was predominately occupied by elderly residents.
Fatalities and Injuries	Four residents died in the Westview Towers fire and there were fifty-four injuries. No one died in the Council Tower Apartments fire, but there were sixty-eight injuries. One firefighter will be unable to return to work as a result of injuries sustained while fighting the fire.
Origin and Cause	The cause of both fires is undetermined, but it is probable that both fires were the result of smoking. The Westview fire began in the living room of a three-room apartment, while the Council Tower Apartments fire began in the bedroom of a three-room apartment.
Fire Protection Systems	Both buildings were equipped with standpipe systems that were supplemented by a fire pump. Neither building was fully equipped with an automatic fire sprinkler system.
Smoke Travel	The absence of pressurized smoke towers contributed to the upward spread of smoke in both buildings and resulted in numerous injuries and two fatalities in the Westview Towers fire.
Access	Access was limited at both buildings and neither was accessible by an aerial device.
Fire Alarm Systems	Hardwired smoke detectors, pull stations, and audible/visual alarm devices were present in the interior hallways of both buildings, which were monitored by a private alarm company. The Council Tower Apartments had a public address system and Westview Towers did not. The apartments in both buildings were equipped with battery-operated smoke detectors.
Smoke Detectors	The residents in the apartment of origin in both fires were using medical oxygen and had a number of cylinders in their apartments. In both instances, the release of the stored oxygen intensified the fire and contributed to fire spread.

The Westview Towers
NORTH BERGEN, NEW JERSEY
AUGUST 9, 1998

Local Contacts: Lt. Thomas Irving, Inspector
                   Battalion Chief Charles Severino
                   North Bergen Fire Department
                   6102 Tonnelle Avenue
                   North Bergen, NJ 07047
                   (201) 392-2162

                   Captain Michael A. Macari, EMT-D
                   North Bergen Emergency Medical Service
                   P.O. Box 873
                   North Bergen, NJ 07047
                   (201) 758-1990

THE BUILDING

The Westview Towers Building, located at 6115 Granton Avenue in North Bergen, New Jersey, was built in 1976. The twenty-two-story building is located in a working class neighborhood with narrow streets that includes a number of buildings that house the elderly. Westview Towers is approximately 230 feet tall and was built of fire resistive construction. There were 296 apartments in the building and the building was fully occupied at the time of the fire. The majority of the 400 residents were elderly. Jack Pomerane owned the building at the time of the fire and the complex is a HUD subsidized housing project.

The main entrance faces to the east and the front door opens onto a sidewalk, which leads up to Granton Avenue. The first floor contains a bank of elevators at the center of the lobby as well as offices, meeting rooms, and a number of apartments. The upper floors are exclusively dwelling units. Access to the front of the building is limited as the entrance is situated below the elevation of Granton Avenue and visitors must walk down a number of steps to reach the front door. Landscaping and overhead power lines also further restrict access.

Access to the south side of the building is limited to foot traffic. A parking lot and driveway located on the north side of the building allows vehicle access as well as provides access to the fire department connection for the standpipe system. A multiple-story parking garage is located along the west side of the building and is not accessible by larger fire department apparatus.

The building had wet standpipes in both the north and south stairwells. Both stairwells were enclosed and had openings to the roof. The standpipe system is supplemented by a fire pump, which operated properly during the fire. The fire alarm system consisted of hardwired smoke detectors and pull stations in the common areas and hallways. The apartments were provided with battery-operated smoke detectors. There were smoke barrier doors on each side of the elevators, which close when the fire alarm system is activated and the building was equipped with emergency lighting. The building was not equipped with an automatic fire sprinkler system. Westview Towers had been inspected prior to the fire and had been found to be in compliance with local codes.

The walls in the hallways and the separations between apartments were built of metal studs covered by fire rated gypsum board. The apartment doors were fire rated and the windows of the individual apartments were operable. A number of the apartments, including the apartment of origin, had balconies, which were accessible by a sliding glass door. The floors of the balconies were made of concrete and had an iron safety railing around the perimeter of each balcony.

FIRE DEPARTMENT PROFILE:

North Bergen, New Jersey is located across the Hudson River from New York City. At the time of the fire, the community of approximately 60,000 residents and was protected by the North Bergen Fire Department. The Department consisted of 100 personnel who were deployed on four suppression shifts. On-duty strength was a minimum of twenty firefighters who staffed four engines, and one truck company and each shift was commanded by a Battalion Chief. On January 1, 1999 the department merged with four area departments to become the North Hudson Regional Fire and Rescue Department.

The North Bergen Emergency Medical Service provides emergency medical services. The EMS service is a combination department with both career and volunteer members.

THE FIRE

The fire alarm company reported the fire at the Westview Towers Building at 16:44 hours on Sunday, August 9, 1998. An occupant of Apartment 4E also telephoned 9-1-1 and reported that her couch was on fire. As the fire progressed, numerous calls to 9-1-1 flooded the North Bergen dispatch center with reports of occupants being trapped in their apartments.

The occupant of Apartment 4F provided a statement to the investigators in which she stated that she was at home at the time of the fire and that she saw both victims in their apartment, which was on fire at the time. The door to Apartment 4E was open and she heard someone shouting, "Help! Fire!" She also stated that she saw that the couch, located at the right side of the room, was on fire. She could see both of the occupants and urged them to leave their apartment. Smoke and fire conditions were such that an escape was still possible at that time.

Smoke conditions then began to worsen in the hallway so the witness went back to her apartment, secured a towel from her apartment to cover her face, and went back to Apartment 4E. She could still see both occupants and urged them to crawl below the smoke towards her and to safety. She continued to urge them to escape until heavy smoke conditions forced her to retreat to the safety of her own balcony where she witnessed both of the victims perish. She was later rescued by firefighters.

FIRE DEPARTMENT ARRIVES

Four engines, a truck company, and the Battalion Chief from North Bergen responded on the initial alarm and began to arrive at 16:47 hours. Engine companies laid 300 feet of 4-inch supply line to nearby hydrants and pumped into the standpipe connection with two 2-1/2-inch lines. Attack crews began to advance 1-3/4-inch hand lines to combat the fire on the fourth floor, but found that heavy smoke and heat conditions would not allow them to reach Apartment 4E.

An attempt was made to reach the two occupants of Apartment 4E from the exterior of the building. They were trapped on their balcony by the extreme heat and heavy smoke and firefighters attempted to reach the balcony by raising a 35-foot ground ladder. Due to the height of the building at this point, the 35-foot ladder would only reach the second floor balcony. Firefighters then carried a 14-foot roof ladder to the second floor balcony and attempted to secure the hooks to the railing on the fourth floor in order to reach the victims.

Unfortunately, the firefighters could not see the victims during their attempt to secure the ladder because of the concrete floor of the balcony above them. The 60-year-old female apparently panicked in spite of reported attempts by firefighters and occupants of an adjacent apartment to calm her. They urged her to wait for the firefighters to reach her, but she lunged for the ladder before it had been secured to the railing. The ladder swayed back and forth and the firefighters holding the ladder were unaware of what was happening. Tragically the victim fell off the ladder to her death. The firefighters were also unable to reach the 90-year-old female and she perished on the balcony after the glass door broke and allowed the intense heat and heavy smoke to reach the victim.

Rescue efforts were hampered by numerous occupants who using the stairwells to exit the burning building. The stairwells were not pressurized and smoke filled both stairwells when residents opened the doors from their hallways into the stairwells. The spread of smoke throughout the building resulted in two additional fatalities and contributed to the large number of people that had to be transported to local hospitals. Firefighters discovered the bodies of the occupants of Apartment 10B in the north stairwell between the sixth and seventh floors. The victims were removed to Apartment 6L where resuscitation efforts proved to be unsuccessful. The victims were later identified as a 49-year-old female and her 65-year-old husband.

Because of the magnitude of the fire, the incident commander struck additional alarms for more personnel. Approximately eighty firefighters responded to the incident. Units from West New York, Weehawken, Jersey City, and Hoboken assisted in the effort to bring the fire under control. At least eighteen engine companies and four truck companies responded along with numerous support vehicles.

Efforts to extinguish the fire proved to be very difficult. The fire resistive construction trapped the heat on the fourth floor. The fire burned with sufficient intensity, apparently due to escaping oxygen from the cylinders in Apartment 4E, to cause the wall between Apartment 4E and the hallway to fail. This slowed the firefighters effort to reach the room of origin.

The fire was commanded throughout the incident by the on-duty Battalion Chief and was finally declared to be under control at 19:33 hours, almost four hours after the initial alarm. Firefighters remained at the scene until 01:05 hours the next morning. During the fire the Governor of New Jersey flew to the site in a helicopter to visit with fire officials and to reassure the residents of the building. The Red Cross provided assistance at the scene and shelter to the occupants displaced by the fire.

INJURIES AND FATALITIES

The fire killed four residents and injured fifty-four people. Thirty-two people, including seven emergency responders, were transported to seven local hospitals primarily because of smoke inhalation and respiratory problems. Table One, below, provides a breakdown of the distribution of the injured. Twenty-two firefighters and an undetermined number of residents were also treated at the scene for minor injuries, cuts, burns, heat stress, and related conditions.

Table 1. Distribution of Patients Transported

Hospital	Location	Patients
Palisade Medical Center	North Bergen	9
Christ Hospital	Jersey City	4
Meadowlands Hospital Medical Center	Secaucus	4
Jersey City Medical Center (trauma center)	Jersey City	5
Holy Name Hospital	Teaneck	1
Saint Frances Hospital	Jersey City	4
Unknown		1
Total:		**32**

The North Bergen EMS was dispatched for a fire standby at the Westview Towers at 16:48 hours and the first ambulance arrived on the scene at 16:50 hours. At 16:51 hours, the EMS Supervisor arrived and assumed EMS Command. Based upon the heavy smoke conditions and the presence of two victims on the balcony, the EMS Commander requested that two additional ambulances and one ALS Unit be dispatched to the fire.

By 17:05 hours, ten patients had been triaged and treated and EMS Command reported that the fire was quickly evolving into a mass casualty incident. EMS Command established the following sectors: triage, treatment, transportation, primary staging, secondary staging, Rehab, and a non-patient holding area. At 19:45 hours the EMS Annex to the County's Emergency Operations Plan was activated. A total of thirty-nine agencies (See Appendix C) assisted in the treatment and transportation of the victims.

THE INVESTIGATION

The fire was investigated by the North Bergen Fire Department in conjunction with the Hudson County Prosecutor's Office and the North Bergen Police Department. The Medical Examiner's Office performed the autopsies on the four victims. The investigation revealed that the fire started in the southwest corner of the living room of Apartment 4E. A small aluminum oxygen cylinder ruptured in this area. A large steel oxygen cylinder and a small aluminum cylinder were also found in the kitchen and the relief devices on both cylinders had actuated during the fire.

The presence of an oxygen enriched atmosphere within Apartment 4E contributed to the intense heat build up, the almost complete combustion of the contents of the living room, the heavy spalling of the concrete ceiling which exposed the reinforcement rods, and the failure of the fire separation between the room of origin and the hallway. The fire separation between the bedroom held and the damage to the contents of the bedroom was far less severe.

The investigation ruled out any mechanical or electrical cause and the fire was not found to have been suspicious or incendiary in origin. The fire has been listed as accidental because investigators could not conclusively determine what caused the blaze. It is, however, believed to have been the result of smoking on the part of the younger of the two victims in Apartment 4E.

ANALYSIS

Fortunately, the fire occurred at a time of the day when most people are alert and active. The results might have been far worse had the fire occurred when the majority of the occupants of the building were asleep. Investigative reports indicate that the residents most immediately threatened were awake at the time of the fire and were aware that the building was on fire. For reasons unknown, the occupants of the apartment in which the fire began failed to escape even though it was possible to do so in the early stages of the fire according to an eyewitness.

The fire was contained to the room of origin and the adjacent hallway. There was damage to the bedroom of Apartment 4E, but the contents were not completely consumed, as was the case in the living room, because the fire did not destroy the wall between the two rooms. It is also possible that the damage to the room of origin would have been less severe and that the wall between the hallway would have held if the oxygen cylinders had not been present. Furthermore, if the building had been equipped with a functioning automatic fire sprinkler system one or two heads would have in all likelihood actuated and controlled the fire with minimal damage or loss of life. Unfortunately, the building code in effect at the time the building was constructed did not require the installation of an automatic fire sprinkler system.

The absence of a public address system in the building and the lack of knowledge on the part of the occupants about what to do in the event of a fire may have contributed to the deaths of the two victims found in the stairwell. Most occupants in fire resistive buildings are safer if they remain in their room and close the door and seal off the opening at the bottom of the door with a wet towel. The use of the stairwells by escaping occupants allowed toxic levels of smoke to accumulate in the stairwells. If the stairwells had been pressurized this would not have occurred and would have reduced the number of injuries and deaths.

The fire department did not have sufficient resources on duty at the time of the fire to successfully manage an incident of this magnitude in the absence of a functioning automatic fire sprinkler system and the firefighters that responded were challenged by the magnitude of the event. Even though mutual-aid companies responded with additional apparatus and personnel, the firefighters were subjected adverse conditions due to the entrapment of heat and smoke within the fire resistive structure.

The steel cylinder and the lack of flame impingement on the aluminum cylinder in the kitchen prevented the oxygen cylinders from rupturing because the relief devices functioned properly and vented the contents of the cylinders into the atmosphere. The energy released by the rupture of a compressed gas cylinder is significant and has the potential to injure or kill anyone in close proximity of the rupture.

Council Tower Apartments
St. Louis Missouri
October 12, 1998

Local Contacts: Neil Svetanics, Fire Chief
Captain Ron Gronemeyer, Chief Fire Investigator
St. Louis Fire Department
1421 North Jefferson
St. Louis, Missouri 63106
(314) 289-1901

THE BUILDING

The Council Tower Apartments Building is located at 310 South Grand in Saint Louis, Missouri and was built in 1969 to house the elderly. It is part of a three building complex that includes the Council House West Apartment Building, constructed in 1966, and a large parking structure. The apartment building was owned by the Teamsters Joint Council 13 and managed by the Sansone Group.

The Council Tower Apartments Building is twenty-seven stories tall was constructed of fire resistive materials consisting of poured concrete re-enforced with steel. The building has a flat room, a full basement, and contains 150 apartments. There were approximately 160 people living in the building at the time of the fire. The floor plan is similar on floors two through twenty-seven and has a single hallway on each floor that is oriented east to west.

There is enclosed stairwell at each end. The east stairwell could be entered from the lobby, extended from the basement to the roof, and had an exterior exit door located at the ground level. The west stairwell extended from the basement to the roof and had an exterior ground level exit, but could not be entered from the lobby. Two passenger elevators and a freight elevator were located off the north side of the hallway. During a fire, the elevators are programmed to automatically return to the lobby.

The exterior apartment doors were solid core pressed composition with a laminated finish and the interior apartment doors were hollow core. The floors were concrete covered by either carpet or vinyl tile. The ceilings were poured concrete with a mortar acoustic finish. Windows spanned the length of the exterior rooms and were single glazed tempered glass. The interior walls were constructed of gypsum board that was screwed onto metal studs and the walls in the hallway had wallpaper that had been painted. Doors into the stairwells were marked as having a B fire rating, but the apartment entry doors were not labeled.

The building was equipped with a voice evacuation system and speakers were located on the south wall in each hallway. There is emergency lighting and a pull station is located at each stairwell exit. The smoke detectors in the hallways were hard wired and were monitored by an alarm company. A simplex alarm panel that displayed the various alarm zones was located in the lobby near the northwest corner of the building. Each apartment was also equipped with a battery-powered smoke detector.

The Council Tower Apartments Building was equipped with a partial sprinkler system that protected the basement, trash chute, and elevator penthouse located on the top floor. The sprinkler system was inter-connected to the standpipe system. The standpipe system consisted of two risers located at the east and west end of the building. A hose cabinet with a 1-1/2-inch connection, a 1-1/2-inch connection, and a fire extinguisher was located adjacent to the doors at each of the stairwells in the interior of the hallway. The two cabinets were 117 feet apart and the standpipe system was supported by a 1,000 gpm fire pumper powered by a 100 hp electric motor. The fire department connection was located on the west side of the building.

Fire department access to the building was limited. The main entrance to the lobby, located on the south side of the building, was covered to protect the residents from the weather and the cover extended westward over the sidewalk to the Council House Apartments Building. The north side of the building was inaccessible by vehicles and the driveway to the building passed through an over-hang in the Council House Apartments Building. Service vehicles and automobiles often congested the drive and the passageways as can be seen in the photos in Appendix D.

The complex was located in an area congested by heavy traffic and was bordered on the south by I 40, on the West by South Grand, on the North by Forest Park, and on the east by the on and off ramps to the confluence of I 40, Forest Park, and Market Streets. Further, the on-site parking was inadequate to handle the number and size of the vehicles and apparatus that was required to control an incident of the magnitude that took place on October 12, 1998.

FIRE DEPARTMENT PROFILE

Saint Louis is located on the west bank of the Mississippi River and has approximately 351,000 residents. The State legislature recognized Saint Louis as a city in 1822 and the fire department has been fully career since 1857. By 1930, Saint Louis had grown to 800,000 residents and the fire department staffed fifty-six engine companies. Unfortunately, Saint Louis suffered hard economic times during the 1970s and 1980s. The city lost population and many of the fire companies were disbanded due to austerity measures.

Currently, the sixty-two square mile city is protected by thirty fire stations that house thirty engine companies; four truck companies, and two heavy rescue squads. The city is divided into six suppression districts and a seventh district protects Lambert International Airport with two stations.

All of the engine companies run quints staffed by four firefighters. At the time of the fire, fifteen of the engine companies were equipped with 50-foot aerial devices and fifteen ran with 75-foot aerial devices. In November 1998, a bond referendum was approved replaced all thirty of the engines with 75-foot quints. The ladder companies are equipped with 110-foot quints.

The annual budget is approximately $37.8 million and there are 873 personnel. Firefighters work a 52-hour workweek. In 1997, the department responded to 109,922 incidents. There were 30,332 fires, 79,590 EMS calls, and 2,781 false alarms.

THE FIRE

At 09:35 hours on Monday October 12, 1998, the sixty-six year old female occupant of Apartment 2103 telephoned the manager of the Council Tower Apartments Building and informed her that her bed was on fire and requested that she call the fire department. Her actions were in accordance with the instructions for reporting fires previously provided to the residents by management. The manager promptly called 9-1-1 and reported the fire. The alarm company reported the fire at 09:44 hours. By that time the incident commander had struck a third alarm.

After reporting the fire, the elderly resident of Apartment 2103 left her apartment and went down to the lobby. She indicated that she left the door to the hallway open when she vacated her apartment. The fire vented itself by breaking out the apartment windows. Firefighters, however, found the apartment door closed when they reached her apartment.

FIRE DEPARTMENT ARRIVES

Engine 29 was the first company on the scene and arrived only minutes after the initial alarm. The officer reported that forty-foot flames were visible on the north side of the building and were coming out of the windows of the twenty-first floor. Before the fire would be declared under control eight alarms would be struck and would summon over 150 firefighters to the incident, including the Chief of the Department who became the incident commander. Approximately forty pieces of apparatus responded leaving only three of the department's fire companies in service in the entire city during the height of the fire. Empty stations were filled by mutual-aid companies and by off-duty firefighters who staffed reserve apparatus.

The crew from Engine 29 attempted to attack the fire with a 13/4-inch handline connected to the standpipe, but found that they could not make the fire floor because of the intense heat and heavy smoke conditions. Meanwhile, the fire had extended out of the windows of Apartment 2103 and had broken the windows of the apartment immediately above, #2203, and had communicated into that apartment. The fire in Apartment 2203 was actually controlled before the fire in Apartment 2103 could be extinguished.

A search of the fire floor revealed that there were twelve apartments on the twenty-first floor. Only five were occupied at the time of the fire (2103, 2104, 2107, 2108, and 2111). The resident of Apartment 2108 was apparently not at home at the time of the fire and as previously stated, the resident of Apartment 2103 escaped to the lobby. The sixty-nine year old male occupant of Apartment 2111 remained in his apartment during the fire and his apartment only suffered minor soot damage. The eighty-year-old female occupant of Apartment 2107 also remained in her apartment during the fire. The seventy-year-old male occupant of Apartment 2104 had to be removed from his apartment by firefighters. When the oxygen cylinders ruptured in Apartment 2013 the separation wall was damaged and allowed the products of combustion to render adjoining Apartment 2104 untenable.

Firefighters and rescue workers evacuated the remainder of the building and used both stairwells and the elevators to remove approximately 160 residents. The stairwells were not pressurized and allowed smoke to seep into the stairwells.

INJURIES

Ten residents required transportation to area hospitals, but none were seriously injured. Three firefighters were also transported. An additional thirty residents and approximately twenty-five firefighters were treated at the scene for minor injuries, but were not transported. EMS service in Saint Louis is provided by the Fire Department.

The forty-seven year old Captain of Engine Company 17 was the most seriously injured. He ran out of air and was trapped on the fire floor. His fellow firefighters located him in the room leading to the trash chute. His SCBA cylinder was empty and he had suffered burns to his respiratory tract. He was wearing a PASS device, but due to the confusion during the fire it is unclear if the device was turned on, or if it operated properly. The Captain was unconscious when he was found by his fellow firefighters and his SCBA was entangled in the wiring of the alarm system. The alarm system had been retrofitted and plastic molding located near the ceiling of the hallway concealed the wiring. The molding failed due to the intense heat and the wires sagged down into the hallway. The Captain suffered severe burns to his respiratory system and was permanently disabled.

The incident was investigated by NIOSH (Report 98F-26 11/23/99). A summary of their recommendations is as follows:

- Ensure that all SOPs regarding highrise firefighting operations are followed.

- Ensure that incident command always maintains close accountability to monitor the location of all firefighters on the fireground.

- Ensure that all officers and firefighters wear and use a PASS device when involved in firefighting or other hazardous duty.

- Ensure that a Rapid Intervention Team is in place before conditions become unsafe.

- Develop and implement a written respirator maintenance program for all respiratory protective equipment used by firefighters.

- Ensure that firefighters entering IDLH atmospheres have fully charged air tanks on their SCBAs.

- Ensure that at least four fighters are on the scene before initiating interior firefighting operations at a structural fire-two in, two out.

- Consider using a rope attached to a permanent object or placing a bright, narrow-beamed light at the entry portal to a structural fire to assist lost or disoriented firefighters in emergency escape.

Ensure that procedures are established to record fireground radio communications.

THE INVESTIGATION

The fire was investigated by the Saint Louis Fire Prevention Bureau and the Saint Louis Police Department. The investigation revealed that nearly all of the contents in the three rooms that comprised Apartment 2103 were totally consumed by the intensity of the fire. The partition wall that separated the living room and the bedroom was totally destroyed, leaving only the metal studs and the electric wires. All of the windows were also destroyed. The east wall, which separated Apartment 2103 from Apartment 2104 also sustained significant damage and had allowed smoke to penetrate into Apartment 2104.

The damage to the partition wall between the two apartments occurred when the aluminum oxygen cylinders in Apartment 2103 ruptured. The remains of three aluminum oxygen cylinders were recovered from the northeast corner of the living room. The release of the oxygen into the burning apartment apparently intensified the rate of combustion and may account for the heavy destruction of the contents and walls. The intensity of the combustion process also contributed to the heat buildup in the hallway and hindered the firefighters in their efforts to extinguish the blaze.

Investigators determined that the area of origin was in the bedroom at the east wall on the north side of the room. Due to the extensive damage the investigators were unable to determine with any degree of certainty the exact cause or ignition source and have classified the fire as "open". The occupant of Apartment 2103 was known to smoke in spite of suffering from chronic bronchitis and emphysema. Investigators collected an ashtray with a cigarette butt in it along with plastic oxygen tubing and a plastic bag as evidence. The items were found in the kitchen sink.

No violations had been recorded during the last inspection according to Fire Department sources.

ANALYSIS

If the building had been equipped with a functioning automatic fire sprinkler system, a single fire company could have handled the fire. Their primary function would have been to mop up the water and restore the sprinkler system. Instead, it took almost the entire on-duty force of a large metropolitan fire department to extinguish the blaze. Fortunately, the fire occurred at a time when most of the residents were awake. The results might have been far worse if the fire had occurred late at night when the majority of the residents would have been asleep.

The incident once again demonstrated the value of fire resistive construction in containing the spread of a fire. The fire was confined to the apartment of origin on the twenty-first floor, but spread to the twenty-second floor through the windows. Glass is not a fire barrier. The fire did not extend beyond Apartment 2203, however, due to the fire resistive construction and the quick action by firefighters.

The incident also demonstrated that it is possible to leave occupants in place in a highrise building constructed of fire resistive materials. There was a significant fire on the twenty-first and twenty-second floors. Nevertheless, several occupants of the twenty-first floor escaped harm by remaining in their apartments.

Unfortunately, the fire also demonstrated the hazards of smoking and of storing oxygen in a residential structure. Firefighters should always assume, until proven otherwise, that any dwelling that houses an elderly or ill resident contains oxygen cylinders. Most of these cylinders will be constructed of aluminum due to their light weight. Aluminum cylinders are susceptible to failure in a fire. At a minimum, cylinder venting will increase the rate of combustion and generate more heat

LESSONS LEARNED

1. **The presence of medical oxygen within a dwelling poses a potential hazard to firefighters.**

   Oxygen is essential for life. Many people who suffer from an acute illness that affects their respiratory system must supplement the amount of oxygen that they breathe. Therefore, these individuals maintain a supply of oxygen within their homes. To assure their mobility, the oxygen supply is usually kept in small cylinders. These cylinders can be made of either aluminum or steel. Aluminum, however, is preferably due to its lighter weight.

   The number of people who rely on supplemental oxygen will no doubt increase as the aged become a larger percentage of the general population and as medical technology expands to sustain the lives of more and more people who previously would not have survived their illnesses. Therefore, there will always be a potential for one or more oxygen cylinders to be present if a fire occurs in the residence of an elderly person.

Firefighters should acquaint themselves with the hazards associated with the presence of medical oxygen. If the stored oxygen is released during a fire, the oxygen will enrich the atmosphere and the rate of combustion will increase and intensify. There is also the potential for a catastrophic failure of an oxygen cylinder, which could injure or kill anyone near the cylinder when it failed.

Eliminating the hazard is not an option. Managing the hazard, however, is an option and the key to proper management is prevention. Fire departments should provide public education programs, which target this special needs population and acquaint the people who use supplemental medical oxygen with the hazards associated with their use of oxygen and the methods to be followed to mitigate the associated dangers. Particular emphasis should be placed upon the safe handling and storage of oxygen cylinders.

Public Education efforts should include acquainting firefighters with the locations of these special needs residents. If possible, a sign or placard should be placed at a conspicuous location alerting firefighters that medical oxygen was in use at that location.

2. **Highrise buildings should be equipped with automatic fire sprinkler systems.**

Automatic fire sprinkler systems have consistently proven to be the most effective form of protection for life and property. Both fires demonstrated the tragic results that can occur when an automatic fire sprinkler system is not present in a highrise building. Fire officials in both jurisdictions believe that the fires would have been relatively minor incidents had sprinklers been in place. Instead, four people perished and a large number of people were injured. A number of residents were displaced by the fires and both buildings sustained major damage.

3. **Fire resistive construction is very effective in limiting the spread of fires beyond the point of origin.**

The fire in the Westview Towers Building was confined to the apartment of origin and the adjacent hallway. A concrete overhang created by the presence of the balcony landing prevented the upward spread of the fire via the exterior of the building. The fire in the Council Tower Apartments Building was confined to the room of origin on the twenty-first floor, but spread to the twenty-second floor via the windows. Nevertheless, the firefighters were able to contain the fire on the twenty-second floor to a single apartment because of compartmentation.

Unfortunately, fire resistive construction can trap heat and smoke and can potentially make the job of extinguishment more difficult. The firefighters in both fires endured severe heat and smoke conditions, which prolonged their extinguishment efforts and no doubt contributed to the number of injuries.

4. **Large scale incidents in highrise buildings are labor intensive and require effective pre-incident plans and mass casualty procedures.**

The North Bergen fire required the services of eighty firefighters to extinguish because of the size of the building and the number of occupants in the building. In Saint Louis, more than 150 firefighters were needed to bring the fire under control. Both of these incidents represent a significant commitment of personnel and equipment. Many fire departments would be hard pressed to muster the resources required to handle an incident of this magnitude.

A pre-incident plan is one of the best ways to identify the resources that will be required to manage an incident in a given target hazard. The plan will help to identify the potential life hazard in the occupancy and well as provide information on the built-in fire protection within a structure. The amount and types of resources will be required to manage an incident should also be included in the plan.

Pre-incident planning saves time can help mitigate many problems before they occur. A pre-incident plan can also be used to identify an occupancy that houses someone who must use medical oxygen to sustain his or her life and can assist firefighters with familiarizing themselves with the location and quantity of oxygen that is being used and stored at these locations.

A mass casualty procedure or plan is also an effective tool for managing an event with a significant number of injuries and/or fatalities. The planning process allows a fire or EMS agency to identify the amount and types of resources that may be necessary to manage an incident and also allows medical facilities to plan for such an event.

5. **Large-scale incidents demand the use of an incident management system and an accountability system.**

A large-scale incident with multiple injuries or fatalities will require a significant commitment of resources and often appear chaotic in the initial stages of the incident. A sound incident management system is essential to effectively and safely manage such events. The presence of a large number of firefighters also increases the potential for a firefighter to become a casualty. An effective accountability system can reduce the potential for that.

6. **Public Education programs designed for the elderly should specifically advise them to not attempt to fight a fire, but rather to safely exit the building and to call 9-1-1 from a safe location.**

According to an eyewitness account, the two residents of the apartment in which the fire started at the Westview Towers could have safely exited their apartment and would have lived had they done so. Unfortunately, they remained in their apartment and attempted to extinguish the fire. By the time they realized that they could not extinguish the fire, it was too late. The fire had increased in size and severity and blocked their only means of escape.

Unfortunately, this is not an uncommon occurrence. Public Educators should always instruct everyone, particularly the elderly, that they should quickly exit their dwelling if a fire should occur. They should be told to go to a safe place and call 9-1-1 to report the fire. At no time should they attempt to extinguish the fire, particularly before the fire has been reported. The instruction should include information on how quickly a fire will grow and spread, reducing their chances for escape.

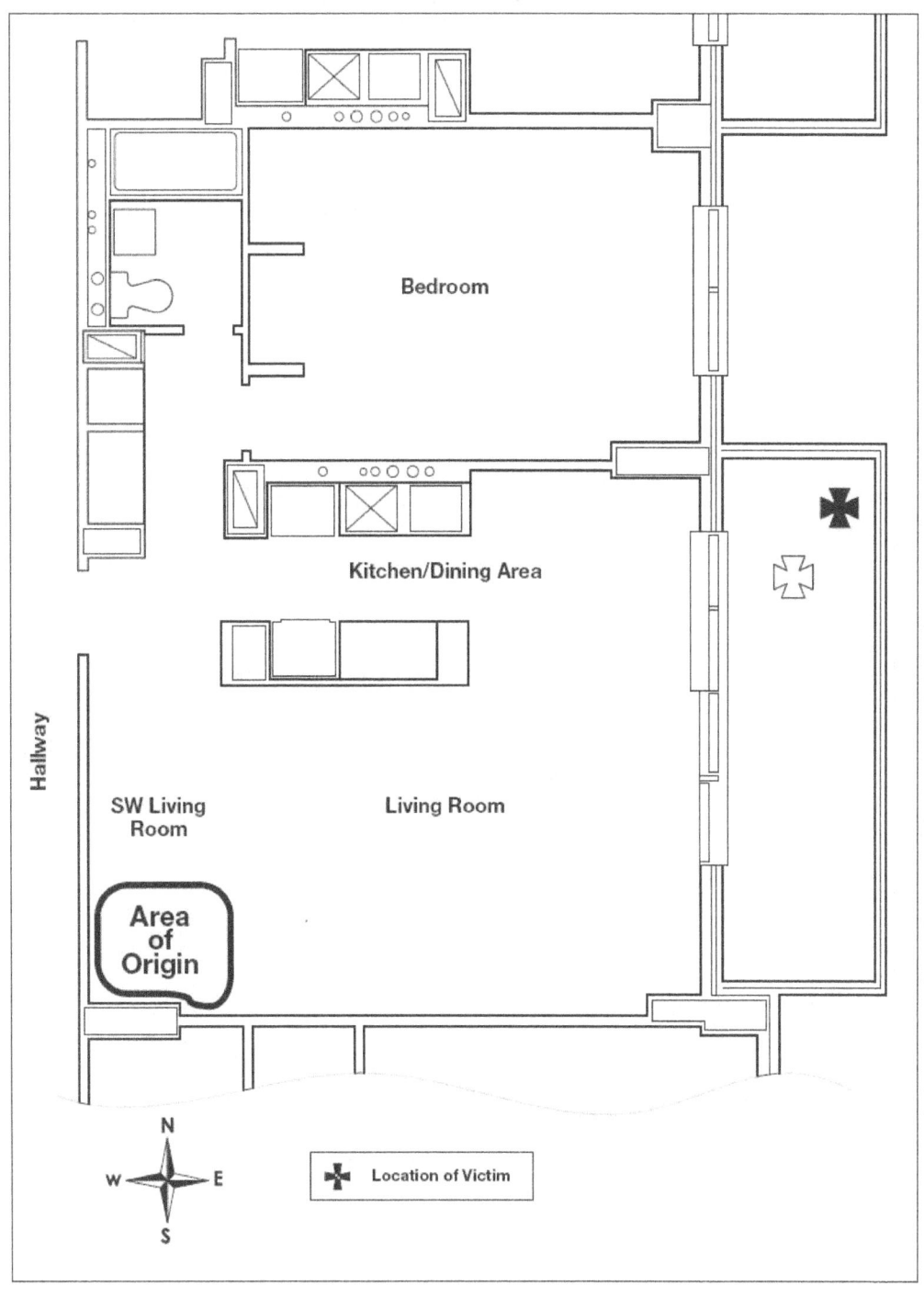

Diagram of the Westview Towers Fire

APPENDIX B

Photographs of the Westview Towers Fire

Photographs by John Lee Cook

Photo Log:

1. Westview Towers – Balcony of Apartment 4E

2. Westview Towers front entrance

North Bergen Fire Department Photos:

None provided due to litigation

Appendix B (continued)

Photo 1. Westview Towers – Balcony of Apartment 4E.

Appendix B (continued)

2.   Westview Towers front entrance

Photo 2. Westview Towers front entrance.

APPENDIX C

List of EMS Agencies That Responded to the Westview Towers Fire

AMB-U-CAR

American Medical Response

Bergenfield

Bogota

Cliffside Park

Clara Mass

Dumont

East Rutherford

Edgewater

Englewood

Englewood Hospital

Fairlawn

Fort Lee

Glenrock

Hasbrouck Heights

Hackensack

Hackensack Hospital

HO-HO-KUS

Jersey City Medical Center

Kearny

Leonia

Little Ferry

Maywood

Moohachie

New Milford

New Jersey First Aid Council

Rochelle Park

Ridgefield

Ridgefield Park

Ridgewood

Rutherford

Saddle Brook

Teaneck

Union City

University Hospital Newark

Wallington

Washington Township

Weehawken

West New York

APPENDIX D

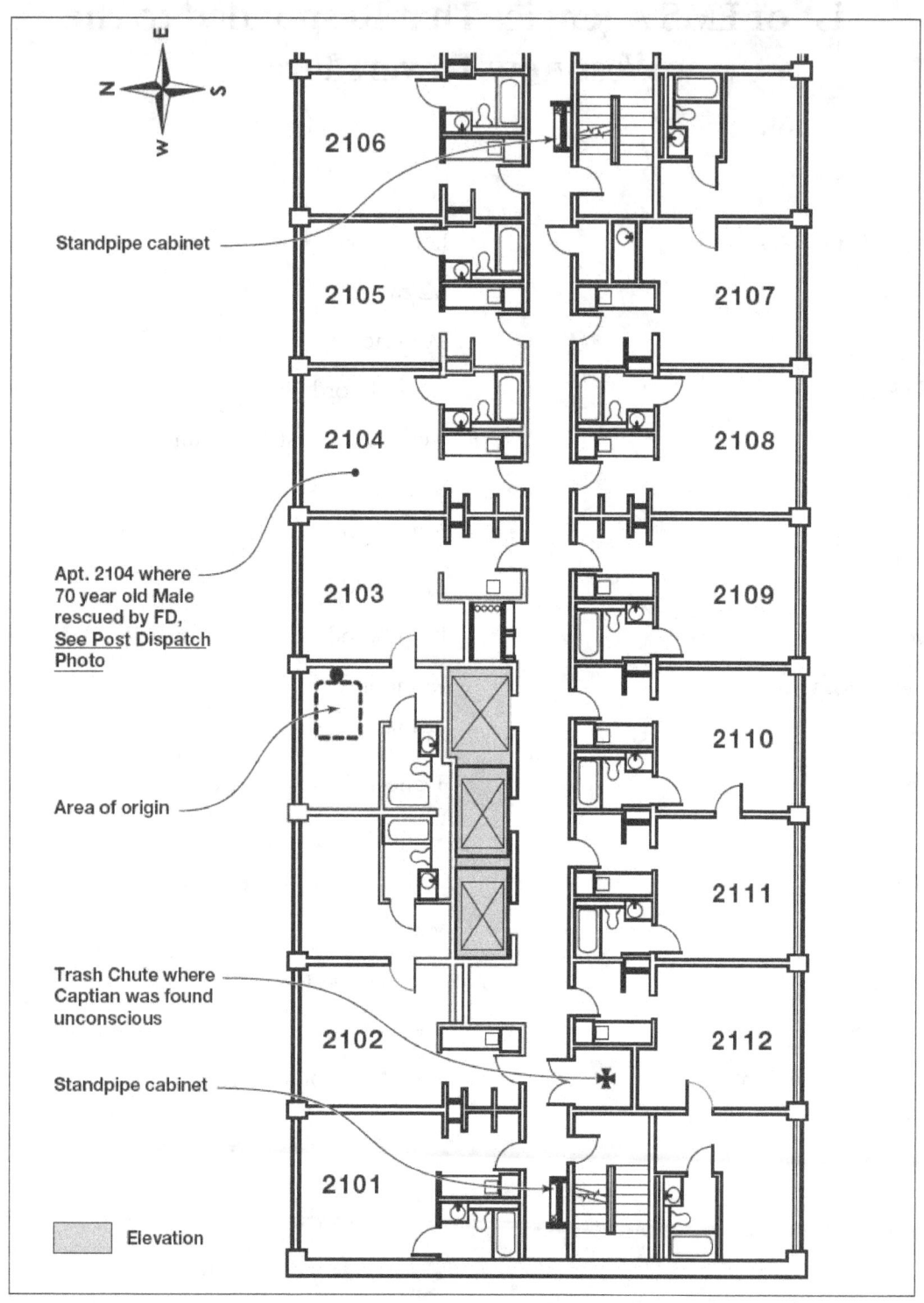

Diagram of the Council Tower Apartment Fire

APPENDIX E

Photographs of the Council Towers Apartments Fire

Photographs 1 through 6: St. Louis Fire Department

1. Ruptured aluminum oxygen cylinder

2. Flames and smoke coming from Apartment 2103

3. Wiring to fire alarm system in the hallway of the 21st floor. (This is the wiring that the Captain became entangled in.)

4. Another shot of the flames on the 21st and 22nd floor

5. Tangled wiring in trash chute where the Captain was found unconscious

6. The interior of Apartment 2103. The area of origin is behind the missing wall in the room where the firefighter is pictured.

Photographs 7 through 9: J. L. Cook

7. Council Towers Apartment North Side. Area of fire involvement located seven rows of windows down from the top. Windows have been replaced.

8. Council Tower Apartment front entrance. (Note access partially obstructed by parked service vehicles.)

9. Council Tower Apartments--View from southeast side showing basement entrance (photo-overlap)

Photograph 10: St. Louis Post Dispatch

10. Occupant of Apartment 2104 awaits rescue.

Appendix E (continued)

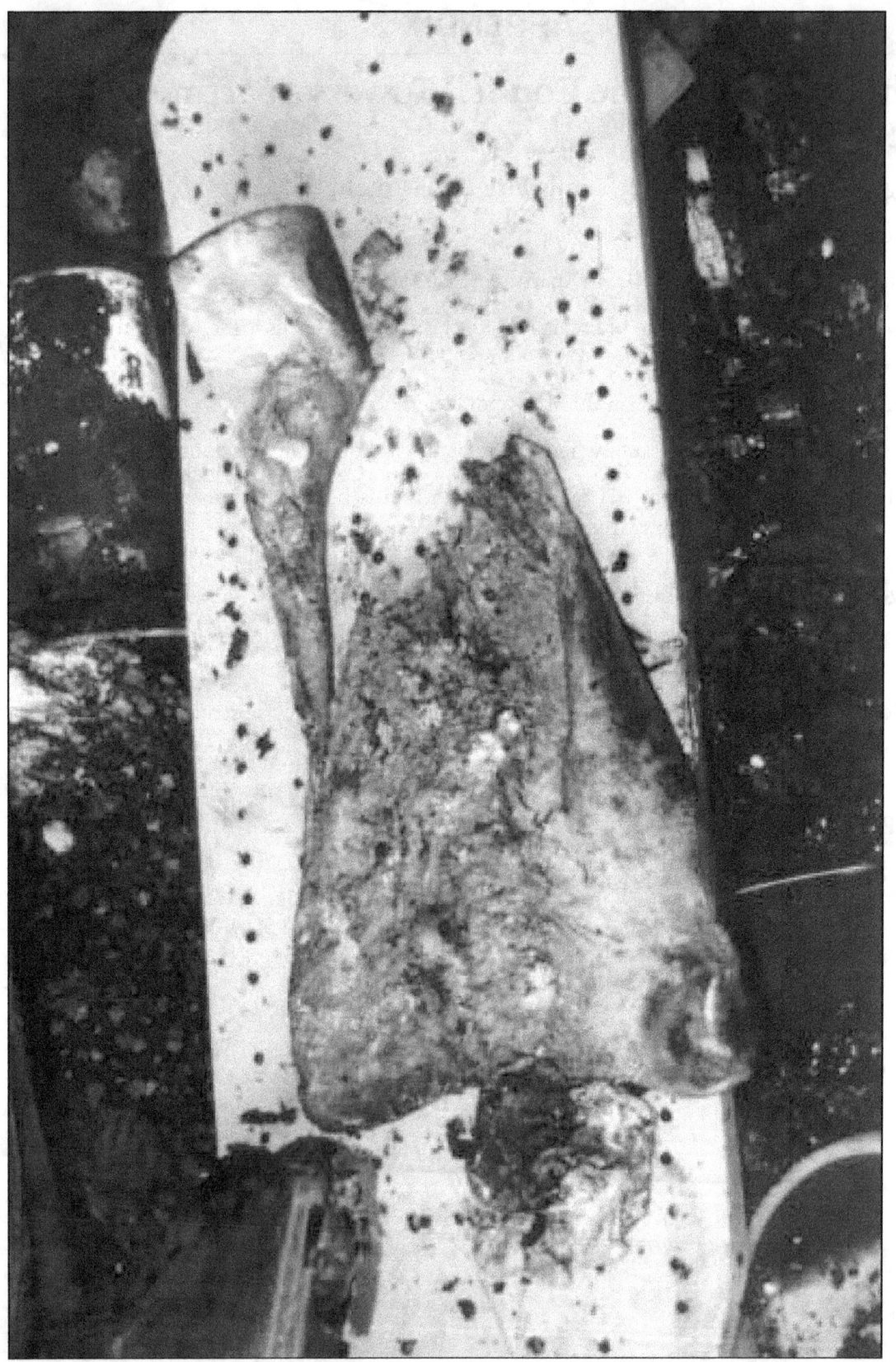

Photo 1. Ruptured aluminum oxygen cylinder

Appendix E (continued)

Photo 2. Flames and smoke coming from Apartment 2103

Appendix E (continued)

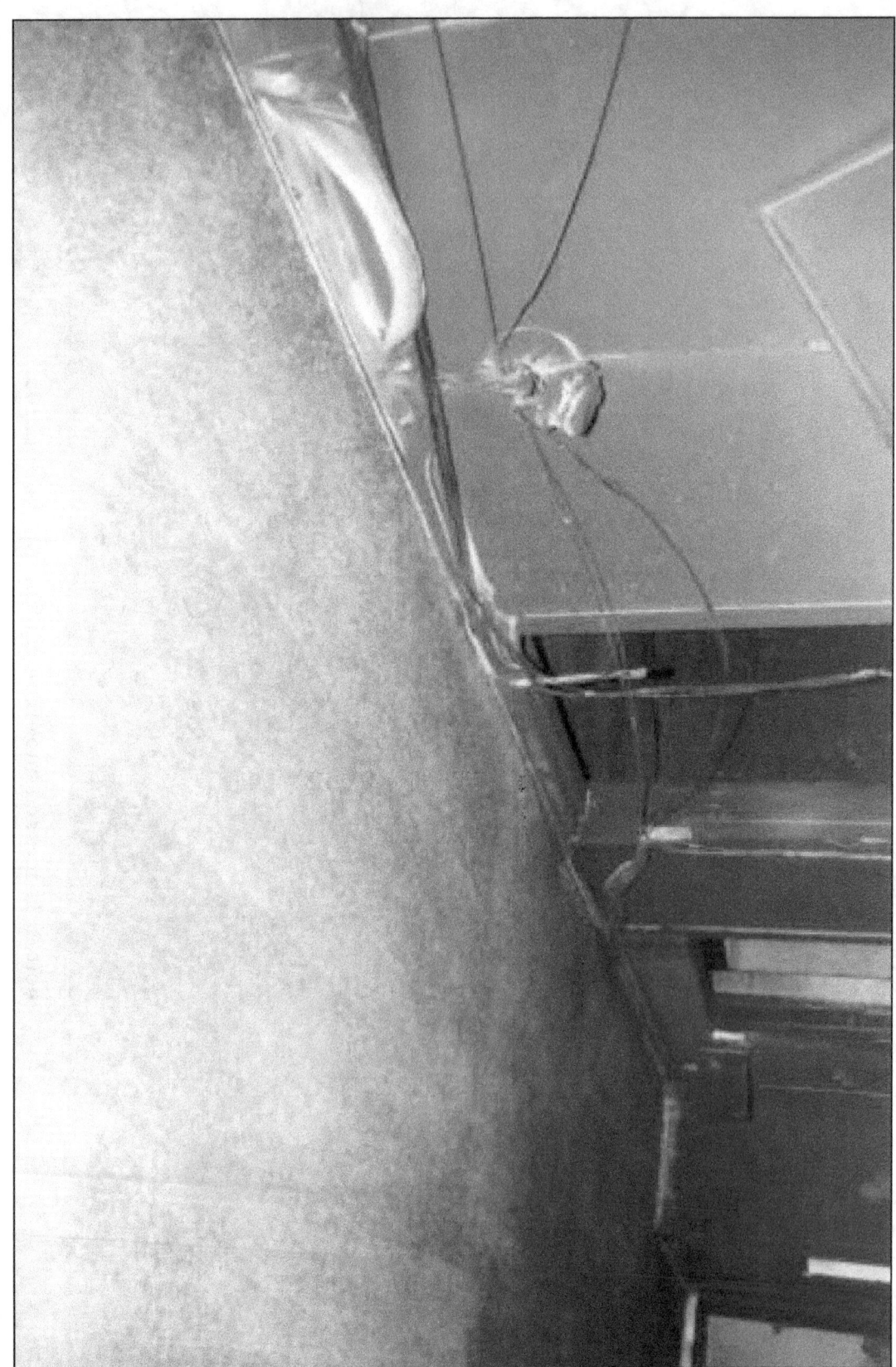

Photo 3 Wiring to fire alarm system in the hallway of the 21st floor (This is the wiring that the Captain became entangled in.)

Appendix E (continued)

Photo 4. Another shot of the flames on the 21st and 22 floor

Appendix E (continued)

Photo 5. Tangled wiring in trash chute where the Captain was found unconscious

Appendix E (continued)

6. Photo 6A. The interior of Apartment 2103. The area of origin is behind the room where the missing wall in the room where the firefighter is pictured.

Appendix E (continued)

Photo 7. Council Towers Apartment North Side. Area of fire involvement located seven rows of windows down from the top. Windows have been replaced.

Appendix E (continued)

Photo 8. Council Tower Apartment front entrance. (Note access partially obstructed by parked service vehicles.)

Appendix E (continued)

**Photo 9. Council Tower Apartments--View from southeast side showing
basement entrance (photo-overlap)**

Appendix E (continued)

Photo 10. Occupant of Apartment 2104 awaits rescue.

APPENDIX F

Medical Oxygen Cylinders

It is recommended that the reader consult Report 107, Fires Involving Medical Oxygen Equipment March 1999, for more information on the impact of medical oxygen on a fire. The chief hazard associated with the presence of medical oxygen within a room or structure is the potential for the release of the oxygen into the atmosphere. Should this occur, the atmosphere might become enriched with oxygen, thus causing the oxygen level to exceed 21% by volume.

As the percentage of oxygen increases above 21%, the fire hazard of most materials change making them easier to ignite and subsequent fires more intense. Even materials, which are difficult to burn or are even noncombustible in air can burn readily and intensely in oxygen enriched atmospheres (100% or less). The amount of oxygen flowing, the size of the room or space, and the degree of confinement will determine the extent of the hazard.

Pure oxygen accelerates combustion resulting in higher flame temperatures than would be associated with fires burning in ambient air. Combustible material exposed to an oxygen-enriched atmosphere lowers the material's ignition temperature. The resulting fires are very intense with rapid combustion and the release of large amounts of energy. If the fire is in a confined space, the rapid temperature rise and increase in pressure often results in the fire penetrating the walls of the enclosure quite quickly. The fire will involve all of the combustible materials in the room sooner than would be normally expected.

Oxygen is a colorless, odorless, and tasteless gas at ordinary temperatures and constitutes approximately twenty-one percent of the earth's atmosphere. Oxygen is essential for both respiration and for the process of combustion. Combustion cannot continue if the volume of oxygen falls below sixteen percent. Conversely, the presence of an atmosphere that is enriched with oxygen will increase the rate at which the combustion process occurs.

The physical properties of oxygen are provided in Figure F.1, below:

Figure 1. Physical Properties of Oxygen

Boiling Point:	-297 °F	-182.8 °C
Freezing Point:	-361 °F	-218.3 °C
Critical Temperature:	-181 °F	-118.4 °C
Critical Pressure:	730 psi	
Vapor Density:	1.1	
Expansion:	1 volume of liquid oxygen expands to 875 volumes of gas	

Source: W.E. Isman and G.P. Carlson (1980). Hazardous Materials. (Encino, CA: Glencoe Publishing Company).

Greater quantities of a gas may be transported and stored in a given size of container by liquefying the gas. Simply compressing the gas into a container can liquefy many gases. Oxygen, however, has a relatively low boiling point, see Figure F.1 above, and may not be liquefied at ordinary temperatures by any amount of pressure. Therefore, the temperature of oxygen must be reduced below its boiling point through a process known as cryogenic production in order to liquefy the gas. This is an expensive and complicated process.

The medical grade of oxygen used almost exclusively by fire and EMS agencies and by individuals is in a gaseous form and is contained in a high pressure cylinder. The United States Department of Transportation and the Food and Drug Administration regulate the manufacture and content specifications of compressed gas cylinders. Oxygen cylinders are manufactured of both aluminum and steel and the capacity of these cylinders range from 5.3 cubic feet to 250 cubic feet.

Compressed gas cylinders are assigned a sequential alphabetical designation by DOT. The D and E cylinders are the two most frequently used small cylinders for medical oxygen and when fully charged contain fifteen and twenty-five cubic feet (650 liters), respectively. Large cylinders are designed G, H, K, and M. An H cylinder, such as the one found in the apartment in North Bergen, contains 244 cubic feet of oxygen when fully charged. The M cylinder, commonly used by fire departments and rescue squads, contains 3,000 liters of oxygen when fully charged.

A fully charged cylinder contains from 2,000 to 2,200 psi at 70 degrees Fahrenheit. This pressure, however, is too high for medical use and requires the use of a pressure-reducing regulator. Regulators typically reduce pressures to 40-70 psi. The cylinder heads have a pin index, which will only accept a regulator designed for medical use. The flow rates are typically low.

The larger cylinders, such as the H, are constructed of steel and the smaller cylinders, such as the D and E, are available in both steel and aluminum. The small aluminum cylinders are very popular due to their light weight. A relief device, known as a frangible disc, is provided to prevent the over pressurization of the cylinder. Over pressurization can occur if a cylinder is over filled or if a cylinder is subjected to excessive temperature, as is the case in a fire. Gases expand when heated. Therefore, the pressure must be reduced or the cylinder will rupture. The bursting pressure of the frangible disc for C and D cylinders is 3,300 psi and approximately 3,750 psi for H cylinders.

The direct impingement of flames on the surface of a cylinder also poses a significant hazard. A cylinder can absorb more heat if the cylinder contains a liquid. A good illustration of this phenomenon is the process of boiling water on a stove in a saucepan. The flame temperature of the gas is much higher than the melting point of the material that the pan is constructed of. The heat is absorbed and the pan remains intact as long as the pan contains water. Once the water completely evaporates the pan begins to melt.

This same phenomenon occurs in a pressurized gas cylinder. A cylinder will remain intact as long as the pressure relief device functions and the cylinder contains sufficient liquid to absorb the heat. Oxygen, however, is in a gaseous form in a cylinder and will not absorb as much heat. The cylinder will rupture if the flame temperature exceeds the melting point of the material that the cylinder is made of. Steel cylinders have a much higher melting point than aluminum cylinders and have a greater potential to endure the flame impingement. Aluminum cylinders are much more prone to failure due to the relatively low melting point of aluminum. The specific alloy composition will, of course, influence the exact melting point of a cylinder.

If a cylinder ruptures, the cylinder fragments can become shrapnel and pose a serious threat to anyone in the vicinity at the time the cylinder failed. The cylinder can also become a projectile and travel considerable distances. An airborne cylinder has the potential to cause considerable damage to anything or anyone in its path.

www.ingramcontent.com/pod-product-compliance
Lightning Source LLC
Chambersburg PA
CBHW081239170526
45165CB00009B/3120